Welcome to the Deep Sky Stacker edition of my astrophotography guides.

I have run a series of successful astrophotography workshops, covering DSLR astrophotography and Webcam imaging and image processing.

The workshops are available on my Web site:
www.star-gazing.co.uk/Workshops.html

Over time, I have started to put together some of the techniques I use into these handy guides.

These should help the newcomer to using these techniques to become more proficient at using the software and achieving better results.

The workflow outlined in these pages is not the only way to do things. Some people may advise differently, but what follows in this guide is certainly the simplest way I have found to achieve perfectly acceptable results.

I hope you find this guide useful.

If you have look out for my other astrophotography guides:
www.star-gazing.co.uk/Guides.html

Dave Eagle.

www.star-gazing,co.uk

May 2018

D1705046

Guide to Stacking Images in Deep Sky Stacker (DSS).

Contents

Deep Sky Stacker is a free piece of software for processing astrophotographs.

It can stack individual images together. This enables the astrophotographer to take many images and add them together.

The software is able to add images together even if there has been some movement between the time the individual images were taken. Even field rotation can be overcome using this software. This is useful for taking wide field images on a static tripod.

We've all heard that there is a need to use dark frames, flat fields, bias frames and other such embellishments to get astrophotographs to be proud of.

Indeed, the use of these techniques will certainly enhance astronomy images, but as shown, they don't have to be used to get a decent image. When first starting out, there is no need to go to these sorts of lengths. In fact, most of the images I have taken have not had these carried out on them.

I try and keep my image processing using the KISS Method: Keeping it Stupidly Simple.

In this guide I work my way through the processing workflow I use within Deep Sky Stacker, so a really decent image can be extracted using just a number of light frames.

Taking the best images for stacking.

Before we get into the processing itself there are some things to make sure images are taken and saved for optimum stacking in DSS.

- Always set the camera to save images in Raw file format.

- Make sure long exposure noise reduction is turned off.

- Take a number of images all the same.

Use the longest exposure possible. This will always be restricted by three things:

- 1) The accuracy of the tracking. Bad tracking will produce bigger star trails. Deep Sky Stacker is very picky about the shape of stars. Any trailing in the images, must be minimal or the software will not recognise and register the stars.

- 2) The amount of light pollution visible in the image. More light pollution quickly fogs the image. It may be necessary to make shorter exposures or use light pollution filters to get around this.

- 3) The speed of movement of the object of interest on the images. This doesn't affect deep-sky objects, but comets are particularly affected as they can move rapidly across the background stars. We will cover comets more on processing for comets in the second half of this DSS guide.

Once the sub-images have been transferred to the computer, they will need to be further processed. Deep Sky Stacker can stack these images together. This will enable image exposures to be taken at the tracking or light pollution limits and then successfully adding them together. This will produce an image that has all the data of a single image with a very long exposure, but taken in very small chunks. We call these sub-images, or subs for short.

What follows is my guide to the processing work flow that I use for virtually all my night sky and comet images.

Download and install Deep Sky Stacker.

This is freeware and is available from this Web site:

http://deepskystacker.free.fr/english/index.html

This is Windows only software, so will not run on an Apple Mac.

Setting Up Deep Sky Stacker.

This is the process used for deep sky objects and comets.

Once the software has been installed, open DSS Software.

Make sure the most up to date version (Currently 4.1.1 -May 2018) is installed.

Earlier versions of the software may not recognise the most up to date versions of camera raw files. So always make sure that the most up to date version is installed.

This may produce unwanted results in the final image producing strange colours in the final image, making them harder to process.

In the lower window are a series of columns, Path, File, etc...

One useful column is the #Stars column. In the default view this is set way off to the right of the lower frame and is hidden.
That's not very helpful.

Use the scroll bar at the bottom to reveal the #Stars column.

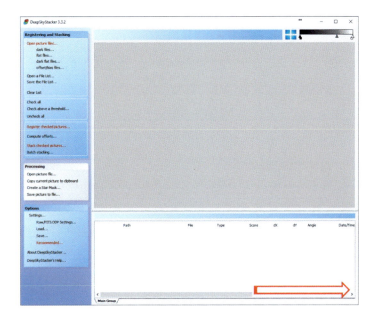

Click and hold on the #Stars tab and drag it further over to the left-hand side of the other tabs, next to the Score column.

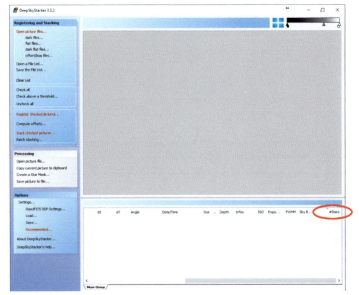

This, along with the Quality tab will be the best guide for judging image quality, so it needs to be visible at all times when using the software.

When happy with the position of the #Stars tab, stacking of the raw images can begin.

Registering Images

Click **Open picture files** at top of the menu.

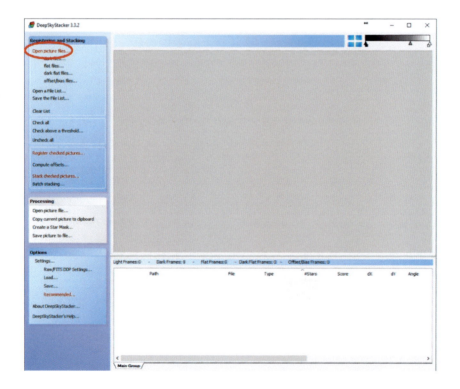

Browse to the folder that contains the images to be stacked.

Highlight all the RAW files required and click **Open.**

In this case I have selected 4 images of the Orion Nebula.

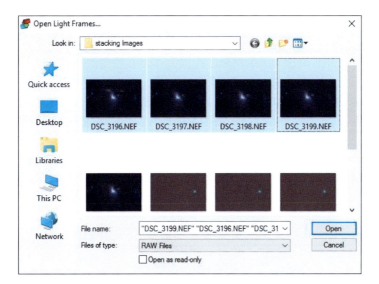

Click **Open** and the selected files will now be visible in the bottom frame.

In practice, more raw images will be used than shown here. The more images used, the better the final stacked image will be. It is all about data. The more images that are stacked, the more data will have been collected.

The more data collected, the smoother and brighter the resulting stacked image will be. A lot of the noise will have smoothed out and the level of pixel brightness across the image will be increased.

The list of selected images will be listed in the lower frame.

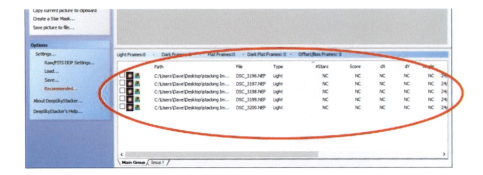

Clicking the mouse on each image in this window will give a preview of the image in the top window.

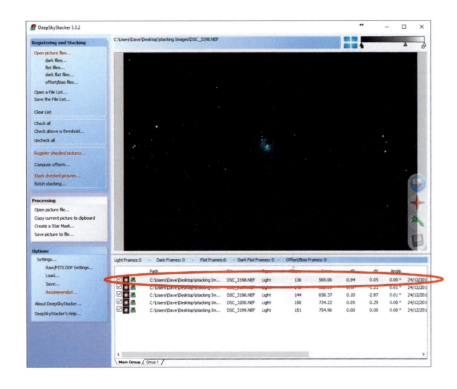

Click on a different image to view it, the new preview will not be visible until the bar at the top showing the image location has turned from red to blue.

Inspect each image to see if the tracking has been accurate, or for the presence of airplane trails, satellites or other interference which could affect the final image.

Click **Check All.**

There should now be a tick in the box against each image as shown below.

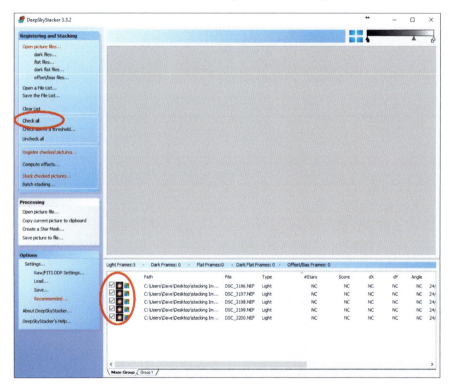

The software is now set, and all should be ready to register all the images.

In the registration process, the software looks for stars in each image and marks the position of each detected star.

Click **Register checked pictures...**

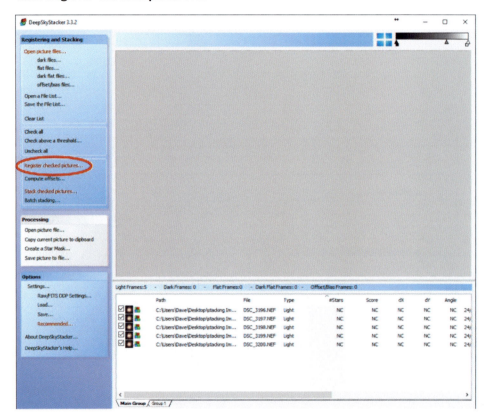

In the window that pops up, there are a couple of changes from the default settings that need to be made before progressing.

Untick the box next to **Register already registered pictures.** (This is optional, depending on whether changes to the sensitivity of stars detected is made or not).

Untick **Stack after registering**.
This will enable some changes to be made before the software stacks the images.

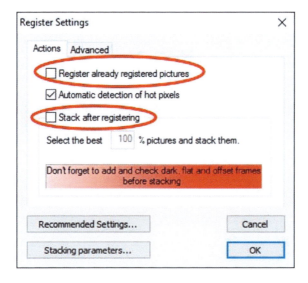

At the top of this window, click the **Advanced** Tab.

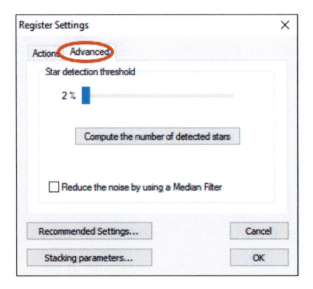

Click the **Compute the number of detected stars** button. Wait while the software examines the first image and looks for stars.

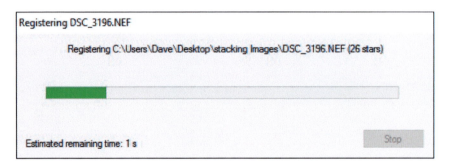

At the end of the process the number of stars it has detected in that image is displayed below the button.

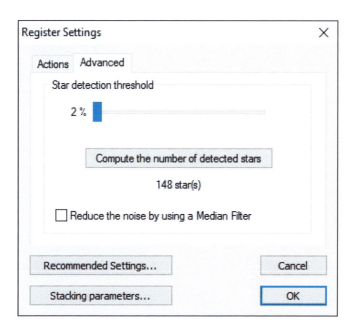

A good number to aim for is around 150 stars.

If too many stars are detected, the software will be doing a lot of work to stack the images later. The stacking process will also take a lot longer.

If the number of stars detected is a lot higher than this, move the slider towards the right to make it less sensitive and re-compute.

Keep adjusting the **Star Detection threshold** and clicking the **Compute the number of detected stars** button until it detects roughly 150 stars. Some practice will give a feel for how much this needs to be changed for each image.

If too few stars are detected, even at 2% threshold, there may be something wrong with the star images. They may be out of focus or trailed too much. DSS is very sensitive about both when detecting stars.

If the number of detected stars is less than 150, even at the 2% threshold, but the star images are OK, stacking may still be possible.

Once happy with the number of stars it can detect, or the software cannot detect any more, click the **OK** button.

A pop up will show that the software is registering the images.

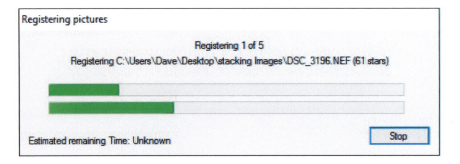

The software is trying to detect stars in the image. It will mark its position in each image and create a record of the stars it has detected in each image.

This information is automatically saved in a text file.

Check in the folder where these raw images are stored, these text files can be seen appearing as the registration process occurs.

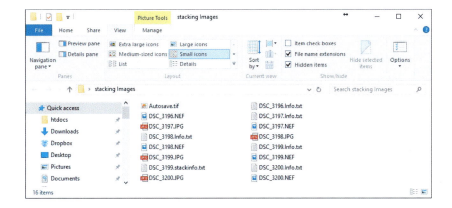

Once DSS has finished registering the images some data is now visible in the lower frame against each image.

The #Stars and Score figures are crucial.
The higher these figures, the better the quality of the sub-image.

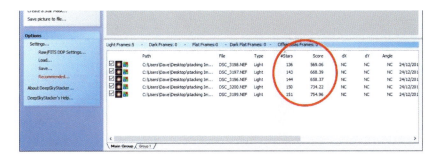

The figures will vary between each image as each image will have a slightly different quality.

If any images are giving a really poor score for either or both of these two counts, it is best not to use the image.

Click on the image again to preview it. A quick view may reveal why the Star Number and Score value is so low. Bad Tracking, camera shake etc.

Stacking Deep Sky Images.

If any images need to be removed from the final stack, untick the box on the left-hand side of the image listing. This image will then be removed from the process and will not become part of the final stacked image.

Once happy that the images selected are OK, click **Stack checked pictures...**

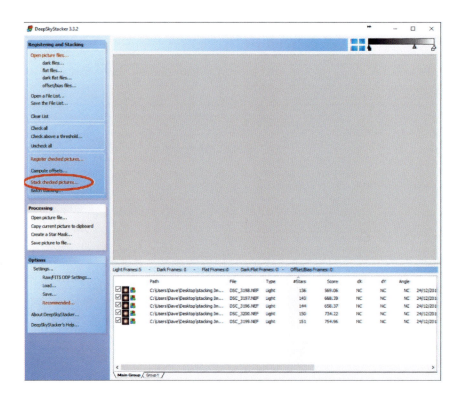

In the window that pops up, click the **Stacking parameters...** button as some changes need to be made before progressing.

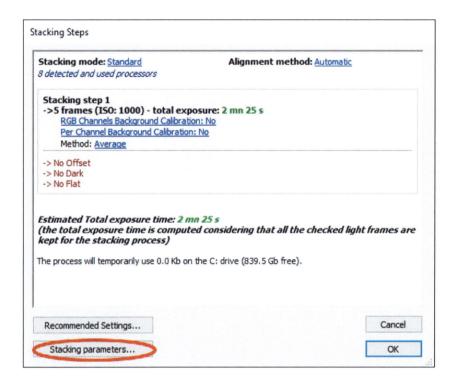

Below are the settings I generally use for quickly processing my images.

Check the settings under each tab and change them to match mine.

Result Tab.

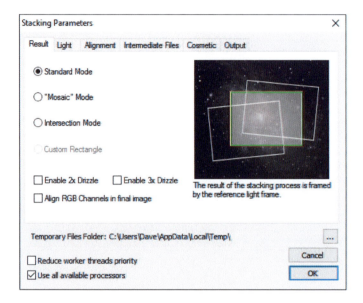

Light Tab.

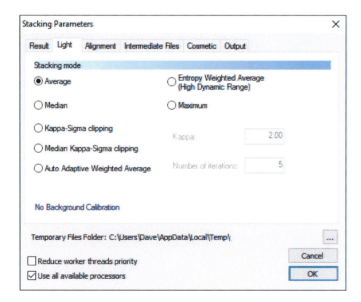

Under the Light tab different settings can be used depending on the outcome required.

I usually use one of these three settings:

Average – The final stack is an average of all the subs.

Kappa-Sigma clipping – Anything not in all the images (airplane trails, meteors, satellites etc.) will not be seen in the final image. Anything out of the average will be blurred out. This will be useful when processing fast moving comet images.

Entropy Weighted Average (High Dynamic Range) – Preserves more or less everything.

Experiment with the different settings to see what works on those images.

Alignment Tab. Leave on Automatic.

Intermediate Files Tab. Make sure TIFF files is selected.

Cosmetic Tab. Leave all unticked.

Output Tab.

Tick **Create Output file.**
Select **Autosave** and tick **Append a number…**

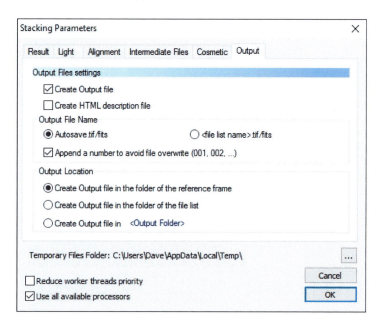

Once all these have been set, everything is now ready to stack the images.

Click the OK button and DSS will now start to stack the images.

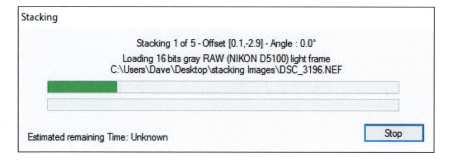

If a mistake is made before starting the stack, do not be fooled into thinking that clicking the **Stop** button will stop the process.

I believe this button has been put in to give the user the feeling of being in control of something. It doesn't work, and the software will carry on stacking regardless.

The stacking process will take quite a long time, especially if a large number of images have been used or being used on a slow computer. This would be a great time to go and make a nice strong cup of tea or coffee.

When the stacking process is finished, the stacked image will be visible in the top frame and some colour information shown in the bottom frame.

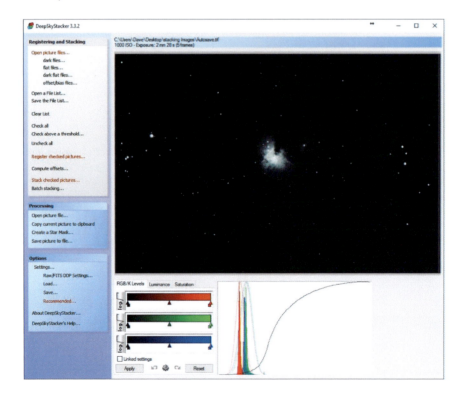

This is what the image looks like compared to the single raw image we saw earlier. Much more of the nebulosity is now visible as the images have been added together.

It will still need further image processing using graphics software to get the best out of it.

Single Sub.   Stacked image.

I never use the colour controls in the lower window to change the levels of the three colours in the image. All my post-stacking image processing is performed outside of DSS on the automatically saved TIF file.

If the same settings as mine, the stacked image will have been automatically saved as Autosave#.tif in the same folder where the original raw images are stored.

One annoying problem that often happens is that the image is saved in a 32-bit format that cannot be opened in some image processing software. To get around this, once the image has been stacked, click **Save picture to file…**

Overwrite the Autosave# file the software has just created or save as a completely new file.

Caution! This save function uses the last folder that a file was saved to. If working on a different set of files stored in a different folder, make sure that the file overwritten is in the current folder. It is very easy to overwrite a file saved previous in a different folder.

Once happy that the image has been saved and the image can be opened in image processing software, click **Clear List.**

This will clear the list of raw files.

If DSS prompts to save the changes as shown below, click **No.**

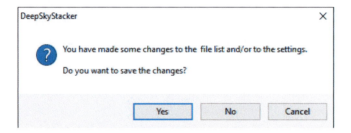

The software has now been cleared to start on the next set of raw images.

This concludes the standard stacking procedure for deep sky objects.

Further processing of the image using a graphics package to get the best out of it.

I cover further processing in my Photoshop Guide.

Stacking Comet Images.

If the images are of a comet, then they need to be treated a bit differently when stacking.

A comet, especially a bright comet, when close to Earth, will change position on each image between exposures, moving, sometimes quite quickly against the more distant background stars. As a result, comet images need to be handled very carefully so detail in the comet is retained. The nearer the comet is to the Earth, the bigger the apparent rate of motion. In the case of fast moving comets, this motion can sometimes restrict the amount of time it can be imaged before the movement blurs the details. So be aware of how fast a comet might be moving in taking the images. Comet 41P, which came very close to Earth in late 2016 and early 2017 was moving so fast, most imagers could only do 30 second subs before the image blurred due to the comets movement, unless they were auto-guiding on the comets nucleus itself.

Capturing Comet images.
When taking comet images, position the comet so that it is framed to get the best features. Don't forget to consider the direction of the tail, to include as much of that in as possible. Placing the nucleus in the centre of the frame is usually not the best position, as the tail may be steaming out of the frame and a lot of detail could be missed. The nucleus to one side with the tail streaming out towards the other side of the frame is ideal.

Once the comet is framed and the best exposure time decided on, take a number of identical images in the same was as shown above for deep sky objects. Take as many images as possible over as long a period as possible. The detail in a comets tail can change markedly, even in a short time, so the aim is to capture as much as possible and reveal those changes.

Once the images have been captured we are now ready to move onto the stacking process.

Registering Comet Images.
The Registration of comet images in DSS is similar to that shown previously for deep sky objects, but with a few changes.

Stacking the images is essentially the same, but this time the images need to be stacked on the moving comet. This ensures that the details in the comet and tail are kept in as sharp detail as possible. We'll need to stack all the images relative to the comets position in each image. To do this we need to tell DSS where the bright central nucleus of the comet is located within each image.

Once the images have been registered, there will be a menu visible in the bottom right-hand side of the image. (I ignored this when discussing the deep sky stacking previously).

Click on the green Comet icon.

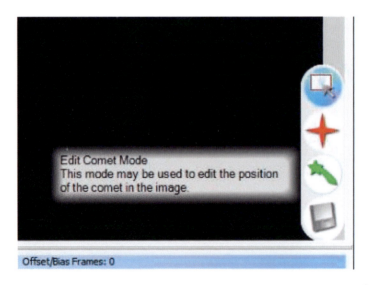

Click on the comet icon, a number of small green circles will appear on the image.

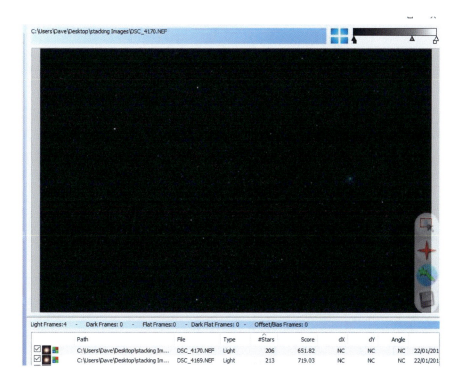

These green circles mark the position of the stars that the star registering process has detected.

Move the mouse over the image window. When the mouse moves over a detected star there is a prompt to set the comet position at this position.

Hover the mouse over the comets position. Use the mouse scroll wheel to zoom into the image for greater accuracy. It can take some getting used to controlling the view this way, but it is worth persevering to get the position of the comet marked as accurately as possible.

Left-click to select the position of the comet in each image.

In many cases the comet may not be bright enough or is too fuzzy to have been detected as a star using the registering process. Don't panic, all is not lost.

In the case of my images of Comet Lovejoy above, the nucleus looks like a big out of focus star towards the right-hand side in the middle of the image. DSS doesn't recognise this as a star, so has not marked its position, despite it being very bright.

To lock the comets position on what DSS "sees" as a blank area of sky, once the mouse is directly over the comets nucleus position, hold down the shift key and left-click on the comets nucleus.

There should now be a mark in the position selected.

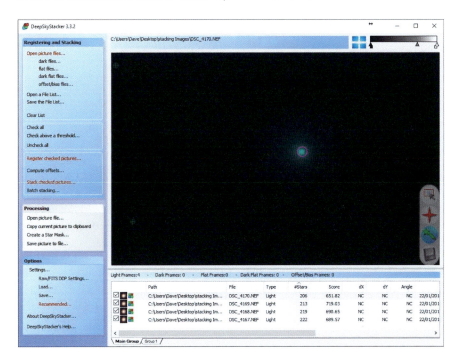

A pink ring will appear around the location of the selected position.

If the location is slightly out, repeat the process until happy with the selected comet position.

It is worthwhile taking time to get this position as accurate as possible, the resulting stacked image will look much better as a result.

Move down and click on the next image and repeat the process, marking the comets position again.

This process needs to be repeated for all the subs, marking the comets position of the comet in each one.

When all the images have had the comets position marked, look under the #Stars column again.

The number of stars detected for each image may have been reduced by one (But not if the comets position was added manually).

The +(C) appearing next to the number also indicates a comets position has been marked in that image.

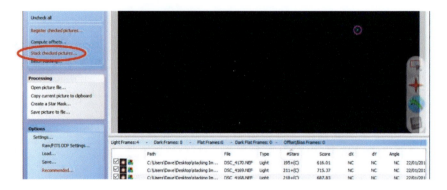

Once the position of a comet has been marked in all the subs, check that all the images have had a comet position marked before progressing by scrolling quickly through the list of files in the bottom pane.

Now all is set nearly to stack the images.

Before moving onto the stacking procedure, think about how many images will be stacked.

Consider: Comets are extremely active, and details can change rapidly.

If so, stacking too many subs over an extended period will result in the tails details being blurred out.

Even when stacked on the comets position, the fast-moving details in the tail will blur as they will have moved during the time the exposures were being taken.

If so, reduce the number of subs stacked to retain as much detail as possible. Experimenting will determine the best strategy to use.

It may also be wise to do the stacking in smaller blocks to produce a series of images to see if the comet tail changes over the total exposure time.

Click **Stack checked pictures...**

In the stacking options window click the **Stacking parameters...** button.

If there is a marked position of the comet in each sub, a new Comet tab should now be visible.

Click the **Comet** tab.

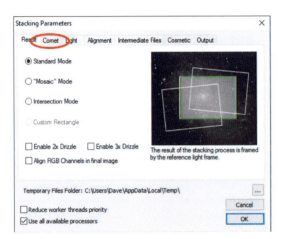

There are now have a selection of comet stacking parameters to choose from.
Select **Comet Stacking.**

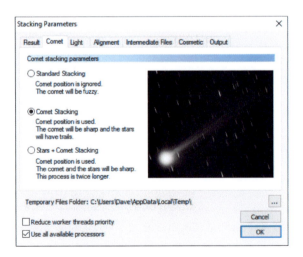

The selections are fairly self-explanatory, but best results for comets are obtained using this Comet Stacking parameter.

Standard Stacking – Ignores the position of the comet and stacks on the position of the stars.

Comet Stacking - Stars will be seen as trails. The length of the star trails shown will depend on the speed of the comet.

Another thing to consider here. If too many images out of the middle of the imaging run are removed and this setting is used, there will be breaks in the resulting star trails in the final image.

Stars + Comet Stacking – Attempts to stack the subs on the Comet and the Background stars. This works best on slower moving comets.

Click on the Light Tab.

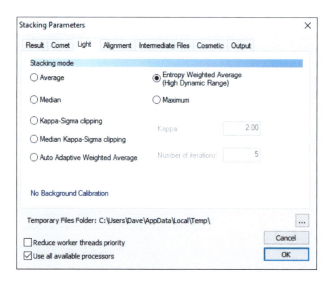

In most of my quick process comet images, I use the setting shown above.

A word of warning. If a full-frame camera is used, the **Entropy Weighted Average** stacking mode may fail due to the amount of processing required.

If either of the Kappa-Sigma clipping modes is sued, the stars will tend to fade away and look a bit ghost-like, especially if the comet is moving quite fast.

Experiment with the settings to see how it works on the images.

Once are happy, click OK.

Click OK on the next window.

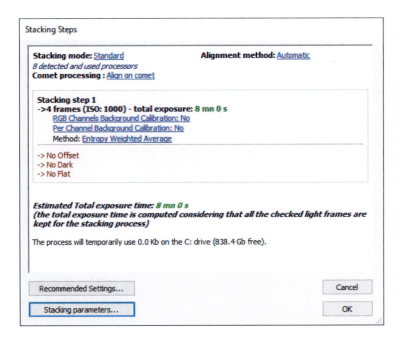

DSS will now start stacking the subs using the comet position marked in each image as the registration point. It will ignore the other detected stars.

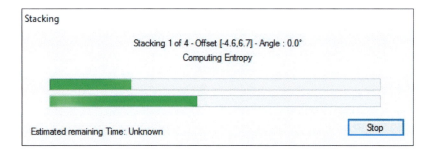

Below is the result of this stack of 4 subs. The stars are only just starting to trail.

When stacking a real image, I would have used a lot more subs and the resulting stars would be very heavily trailed once the images have been stacked. See the resulting image of comet Lovejoy on the cover of this guide.

The comet itself is nicely stacked, showing very little distortion as shown below.

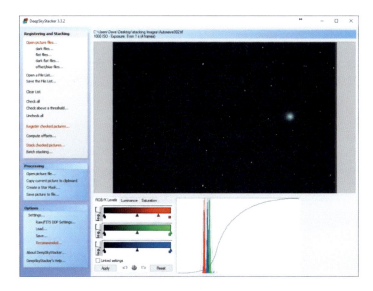

As before, the Autosaved tif image will be saved to the same folder as the raw images.

That stacked tif file will now need to be further processed using image processing software to get the best out of it.

I hope that this guide has built familiarity with the stacking procedure of Deep Sky Stacker and helped get the best out of the images.

If I have made any mistakes in laying out this guide, or something hasn't quite been made clear enough, please let me know so I can improve future editions.

I am also open to correspondence and love to hear from fellow enthusiasts, so please send me a message.

If you have found this guide useful, I also run successful astrophotography imaging hands-on workshops.

For the latest dates, see my Web Page:
www.star-gazing.co.uk/Workshops.html

dave@star-gazing.co.uk

Please follow me on:

My blog: www.star-gazing.co.uk/Blog.html

Facebook: https://www.facebook.com/Eagleseye45

Twitter: https://twitter.com/DaveEagle45

Flickr: https://www.flickr.com/photos/eagleseyeonthesky

Currently available for purchase are two other astrophotography guides:

Guide to Photoshop Astrophotography Image Processing.
Guide to Solar Webcam Imaging.

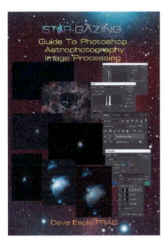 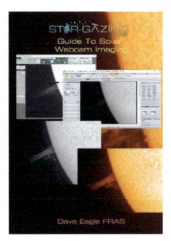

http://www.star-gazing.co.uk/WebPage/astrophotography-guides

Coming Soon: Guide to Imaging the Moon.

Keep Looking Up.

Dave Eagle FRAS.

May 2018.

www.star-gazing.co.uk

Printed in Poland
by Amazon Fulfillment
Poland Sp. z o.o., Wrocław

32297921R00025